THE TEN ESSENTIALS

For Travel in the Outdoors

The Mountaineers

5 4 3
5 4 3 2 1

Published by The Mountaineers
1011 SW Klickitat Way, Seattle, Washington 98134
1-800-553-4453

♻

Manufactured in the United States of America
on recycled paper.

Cover illustrations by Nick Gregoric
Text illustrations by Marge Mueller
Typesetting by The Mountaineers Books

Adapted from *Mountaineering: Freedom of the Hills, 5th
edition*, published by The Mountaineers.

The Mountaineers is a non-profit outdoor activity club
devoted to exploring, studying, preserving, and enjoying
the natural beauty of the outdoors. For membership
information write to The Mountaineers, 300 Third Ave.
West, Seattle, WA 98119.

ISBN 0-89886-374-0

CONTENTS

ESSENTIAL EQUIPMENT

In packing for a wilderness trip, it's a simple matter of take it or leave it. The idea is to take what you need and to leave the rest at home. With thousands of choices widely available in outdoor clothing and equipment, it's no longer a question of how to find what you need, but rather of limiting your load to just the items that will keep you safe, dry, and comfortable.

To strike a balance between too much and too little, monitor what you take with you. After each trip determine what was used, what was really needed for a margin of safety, and what was not needed. As you buy equipment, go for lightweight alternatives if the weight reduction does not jeopardize the item's performance and durability. Whenever possible, look for versatile equipment you can use for several purposes.

If you're new to wilderness travel, you don't have the experience yet to know what works best for you. So don't buy all your basic gear right away. Take it one trip at a time, one purchase at a time.

Whether it's boots, packs, or sleeping bags, wait until you have enough experience to make intelligent decisions before spending your money. Rent, borrow, or improvise during your early outings until you learn what you need and what you don't need. Get advice by talking to seasoned hikers and climbers, by window-shopping at outdoors stores, and by reading mountaineering magazines.

There is a selection of small but critical items that deserve a place in almost every pack. You won't use every one of these items on every trip, but they can be lifesavers in an emergency, insurance against the unexpected.

Exactly how much "insurance" you should carry is a matter of debate. Some respected minimalists argue that weighing down your pack with insurance items causes you to hike or climb slower, making it more likely you'll get caught by a storm or nightfall and be forced to bivouac. "Don't carry bivy gear unless you plan to bivy," they argue.

The majority of hikers and climbers, however, take along carefully selected safety items to survive the unexpected. They sacrifice some speed but argue they will always be around tomorrow to attempt again what they failed to climb today.

The special items most outdoors people believe should always be with you have become known as the Ten Essentials.

The Ten Essentials are:

1. **Flashlight/headlamp, with spare bulbs and batteries**
2. **Map**
3. **Compass**
4. **Extra food**
5. **Extra clothing**
6. **Sunglasses**
7. **First-aid supplies**
8. **Pocket knife**
9. **Matches, in waterproof container**
10. **Fire starter**

Other critical items often join the list of essential equipment, depending on the trip. We will take a look at the Ten Essentials and at a number of these additional items in the sections that follow.

THE TEN ESSENTIALS

1. FLASHLIGHT/HEADLAMP

Headlamps and hand-held flashlights are important enough and temperamental enough to make it worthwhile to invest in only quality equipment. Waterproof flashlights are worth the added expense. They function reliably in all weather, and the contacts or batteries won't corrode even if stored in a moist garage for months.

Few headlamps are waterproof, but some are more water-resistant than others. If you decide that a truly waterproof headlamp is important, start by buying a small waterproof flashlight. Sew a length of wide elastic into a headband, then sew several small retaining loops of thinner elastic onto the headband to hold the flashlight in place. The flashlight will be held on the side of your head, much like a pencil resting on top of your ear.

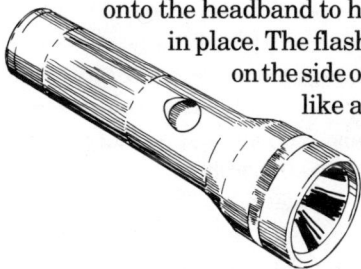

All lights need durable switches that cannot turn on

accidentally in the pack, a common and potentially serious problem. Switches tucked away in a recessed cavity are excellent. So are rotating switches in which the body of the flashlight must be twisted a half turn. If it looks like your light switch could be tripped accidentally, guard against this danger by taping the switch closed or removing the bulb or reversing the batteries. Adjustable focus is an excellent feature available on some lights. The reflector rotates so that the lamp gives flood lighting for chores close at hand or spot lighting for viewing objects far away. Adjustable focus permits maximum use of your light, often letting you see farther and accomplish more than a brighter light lacking this feature.

Flashlight bulbs don't last long, so carry spare bulbs in addition to spare batteries. You don't have to stick with the bulb that comes with your light. If it's a vacuum bulb, you can get a brighter beam with a replacement bulb filled with a gas, such as halogen, krypton, or xenon. These gases allow filaments to burn hotter and brighter than in vacuum bulbs, though they also draw more current (amperage) and shorten battery life.

Most bulbs have their amperage requirement marked on the base. You can get a rough idea of battery life by dividing the bulb's amperage figure into the amp/hours assigned to the batteries. For example, batteries rated at 4 amp/hours will burn about 8 hours with a .5-amp bulb.

It's a good idea to carry bulbs drawing different amperages. Conserve batteries by using low-amperage bulbs for tasks around camp. Switch over to high-amperage bulbs when you need a brighter beam.

Alkaline batteries are the best general-purpose batteries commonly available at mass merchandisers. They pack more energy than cheaper lead-zinc batteries. The major problems with alkalines are that voltage (hence brightness) drops significantly as they discharge, they can't be recharged, and their life is drastically shortened by cold temperatures (they operate at only 10 to 20 percent efficiency at 0 degrees Fahrenheit).

Nickel-cadmium batteries (nicads) can be recharged up to a thousand times, maintain their voltage and brightness throughout most of their discharge, and function well in the cold (about 70 percent efficient at 0 degrees Fahrenheit). However, they don't store as much energy as alkalines. For climbing or remote area travel, look for high-capacity nicads, which pack two to three times the charge of standard nicads and are worth the added expense.

Lithium batteries have twice the voltage of regular batteries, so you'll need to rewire your light to run off half as many batteries. But one lithium cell packs more than twice the amp/hours of two alkalines. The voltage remains almost constant over the life of a lithium battery, and its

9

efficiency at 0 degrees Fahrenheit is nearly the same as at room temperature. They are expensive, however.

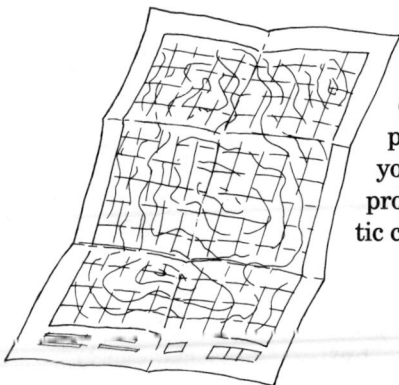

2. MAP

Always carry a detailed topographic map of the area you are visiting, in a protective case or plastic covering.

3. COMPASS

A compass is an essential tool of navigation and route-finding. Learn how to use it.

4. EXTRA FOOD

A one-day supply of extra food is a reasonable emergency stockpile in case you are delayed by foul weather, faulty navigation, injury, or other reasons. The food should require no cooking, be lightweight and easily digestible, and store well for long periods. A combination of jerky, nuts, candy, granola, and dried fruit works well. Some hikers and climbers also bring extra cocoa, dried soup, and tea if a heat source is available.

5. EXTRA CLOTHING

How much extra insulation is necessary for an emergency? The garments used during the active portion of a hike or climb and considered to be your basic clothing outfit are the inner and outer socks, boots, underwear, trousers, shirt, sweater or pile coat, hat, mitts or gloves, and raingear. These garments suffice over a wide range of temperatures and weather if the wearer is active. Extra clothes for an unplanned bivouac are added to these basics according to the season. Ask yourself this question: What do I need to survive the worst conditions I could realistically encounter?

Extra underwear is a valuable addition that weighs little. During a strenuous climb, perspiration can soak the next-to-skin layer and conduct heat away from the core. It is important to get rid of wet garments in contact with the skin before donning thick er insulation, such as down or synthetic coats and an outer shell. Protection for the neck and head can be gained by using long underwear with a zippered turtleneck, and a balaclava. An additional thick hat will retain almost as much warmth as an additional sweater. For the feet, bring an extra pair of heavy socks; for the hands, an extra pair of polyester or pile mitts.

For winter trips and expedition climbing in severe conditions, you'll need more insulation for the torso as well as insulated overpants for the legs. In addition to your rain shell, carry some sort of extra shelter from the rain, such as a plastic tube tent, a Mylar "space blanket" taped into the shape of a bivy sack, or plastic trash-can liners. Carry an ensolite seat pad to reduce heat loss to the snow (some packs can double for this purpose).

Some winter travelers carry a bivouac sack, weighing about a pound, as part of their survival gear and go a little lighter on their insulating layers. It's a good strategy. The sack protects your insulation from foul weather, minimizes the ef-

fects of convection by quieting the air surrounding you, and traps much of the heat escaping from your body so that the temperature of your cocoon rises.

6. SUNGLASSES

Eyes are particularly vulnerable to the brilliance of mountain skies, visible light reflecting off snow, and ultraviolet rays—which at 10,000 feet are 50 percent greater than at sea level. The retinas of unprotected eyes can be easily burned, causing the excruciatingly painful condition known as snowblindness.

Because damage occurs to the eyes before discomfort is felt, it is essential to wear sunglasses to reduce the visible light and invisible ultraviolet rays. Don't let cloudy conditions fool you into leaving your eyes unprotected. Ultraviolet rays penetrate clouds, and the glare of reflected light can cause headaches. Sunglasses should filter 95 to 100 percent of the ultraviolet light. They should also be tinted so that only a fraction of the ambient light is transmitted through the lens to your eyes. For glacier glasses you want a lens with a 5 to 10 percent transmission rate. For more general outdoor use, the lenses

should have about a 20 percent transmission rate. Look in a mirror when trying on sunglasses: if you can easily see your eyes, the lenses are too light. Lens tints should be gray or green if you want the truest color, and yellow if you want visibility in overcast or foggy conditions.

There is little proof that infrared rays (heat-carrying rays) harm the eyes unless you look directly at the sun, but any product that filters out a high percentage of infrared, as most sunglasses do, is added eye insurance.

The frames of sunglasses should have side shields that reduce the light reaching your eyes, yet allow adequate ventilation to prevent fogging. Problems with fogging can be reduced by using one of the many anti-fog lens cleaning products that are available.

Groups should carry at least one pair of spare sunglasses in case a pair is lost or forgotten. You can also improvise eye protection by cutting small slits in an eye cover made of cardboard or cloth.

Anyone who depends on corrective lenses should carry a spare pair of glasses. Normal glasses with clip-on sunglasses can function as one pair, and good prescription sunglasses with extra-dark coatings can be the second pair. Alternatively, this person can carry two normal pairs of prescription glasses plus a well-ventilated pair of ski goggles outfitted with a very dark lens.

Many hikers who need corrective lenses prefer

using contact lenses instead of glasses. Contacts improve visual acuity, don't slide down your nose, don't get water spots, and allow the use of non-prescription sunglasses. Contacts do have some problems, however. Sun and blowing dust and dirt can dry out or irritate your eyes, though good side shields reduce the ill effects. Backcountry conditions make it difficult to clean and maintain the lenses.

7. FIRST-AID SUPPLIES

Don't let a first-aid kit give you a false sense of security. It cures very few ills. Doctors say that in the field there is often little they can do for a serious injury or affliction except initiate basic stabilizing procedures and evacuate the patient. The best course of action is to take the steps necessary to avoid injury or sickness in the first place. Still, you want to be prepared for the unexpected.

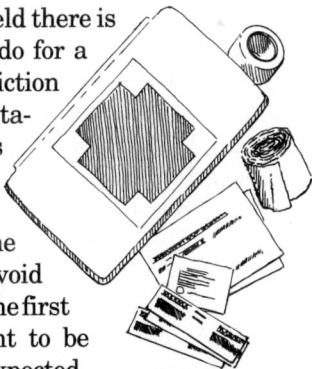

Your first-aid kit should be small, compact, and sturdy, with the contents wrapped in waterproof packaging. Commercial first-aid kits are widely available, though many are inadequate. (Two of the best sources of information on items to include in a first-aid kit are *Medicine for Mountaineering,* by James A. Wilkerson, M.D., and *Mountaineering First Aid,* by Marty Lentz, Steven C. Macdonald, and Jan D. Carline, both published by The Mountaineers.)

The contents of your kit will be items you carry on all trips: Band-Aids or other small adhesive bandages, moleskin, adhesive tape, cleansers, soap, butterfly bandages, scissors, triangular bandages, a carlisle bandage, roller gauze, and gauze pads in various sizes. Carry enough bandages and gauzes to absorb a significant quantity of blood. Severe bleeding wounds are a common backcountry injury, and sterile absorbent material cannot be readily improvised.

In addition to the basics, consider the length and nature of your trip. If you are traveling on a glacier, for example, tree branches won't be available for improvised splints, so a wire ladder splint would be extremely valuable in the event of a fracture. For a wilderness expedition, you may need to consider appropriate prescription medicines.

8. POCKET KNIFE

Knives are so essential for food preparation, fire building, first aid, and even rock climbing that every backcountry traveler needs to carry one. The knife should have two folding blades, a can opener, a combination screwdriver and bottle

opener, scissors, and an awl. The tools and the inside of the casing should be made of stainless steel. Although you may carry a few additional tools on certain trips, the knife is your basic tool kit. A cord attached to the knife and secured to your belt lets you keep the knife in your pocket for ready access without danger of losing it.

9. MATCHES

An emergency supply of matches, stored in a watertight container, should be carried on every trip in addition to the matches or butane lighter used routinely. Storing waterproof or wooden matches and a strip of sandpaper in a film canister makes a good emergency system.

10. FIRE STARTER

Fire starters are indispensable for igniting wet wood or starting a fire quickly in an emergency. They can even be used to warm a cup of water or soup, if you have a metal cup to heat it in. Common fire starters are candles or solid chemical fuel (such as heat tabs or canned heat).

OTHER IMPORTANT ITEMS

WATER AND WATER BOTTLES

Higher areas are often bone dry or frozen solid, so you usually carry the water that is needed to prevent dehydration and maintain energy. A single 1-quart bottle usually suffices, but if it's hot and you sweat a lot, you may need to carry two. Wide-mouthed polyethylene bottles are the most popular because they can be easily refilled with snow or water, and they impart no flavor to the water, unlike aluminum. If you do use an aluminum bottle, be careful not to mix up the water bottle with the bottle that carries fuel for the stove.

Giardiasis, a waterborne intestinal parasite, is a major health concern for alpine travelers. The cyst of the protozoan has tainted many water sources in the wilds, and it takes only one swallow of contaminated water to get sick. The illness takes six to twenty days to manifest itself, with symptoms that include nausea, flatulence, diarrhea, cramps, belching, fever, and dehydration.

Several methods of water purification make water safe to drink by killing or filtering out *Giardia lamblia* and other contaminants. Chemical treatment with tablets of iodine or halazone is effective, and tetraglycine hydroperidide (TGHP) tablets are better yet. Follow directions closely. The required amount of chemical and soaking time increases dramatically with colder water. If you fill your bottle from a glacial stream, you may need to place the bottle in the sun or inside your shirt to raise the water's temperature.

Bringing water to a rolling boil will kill *Giardia* cysts but may not kill other disease-causing bacteria and viruses unless the water boils for nearly 20 minutes. A combination of chemical treatment and a short boil makes water safe.

Water filters are another option. Filters with a pump can strain a quart of water in minutes and do not impart any unpleasant taste to the water. A good water-filtering device provides a method to clean and extend the life of the filter, which is replaceable. Buy only filters with pore sizes smaller than 4/10 micron (.4 micron) to ensure you will filter out *Giardia* cysts, flukes, tapeworms, and protozoa.

SUNBURN PREVENTION

Sunlight at high altitudes has a burning capacity many times greater than at sea level, so much so that it is a threat to both comfort and health. Climbers especially cannot avoid long exposure to the sun so they must reduce the burning ultraviolet rays reaching their bodies by covering their skin with clothing or burn-preventing creams (sunscreens).

Individuals vary widely in natural pigmentation and the amount of protection their exposed skin needs. There is only one rule here: the penalty for underestimating the degree of protection needed is so severe, including the possibility of skin cancer, that you should always protect your skin.

Clothing is by far the best sun protection and is worth the discomfort it causes on a hot day. Skin not covered by clothing should be covered with sunscreen. Most sunscreens use p-amino benzoic acid (PABA), but if your skin is sensitive to PABA there are other chemicals that yield the same sun protection factor (SPF) as PABA creams.

For glacier travel or a snowy environment, get a sunscreen with an SPF of 15 or more. Cover all exposed skin, even if it's shaded, because reflected light can burn the underside of your chin and nose and the insides of your ears. Some sunscreens are advertised as waterproof and will protect you longer than regular products when you are sweating heavily. Regardless of the sunscreen used, however, be sure to reapply it occasionally if you are perspiring.

Actor's grease paint (clown white) or zinc-oxide pastes ensure complete protection, and the grease bases keep them from washing off. One application lasts the entire climb, except where fingers or equipment rub the skin bare. The disadvantages of these creams is that they are messy and so difficult to remove that you may need help from a cold-cream cleanser.

The area around the mouth is particularly susceptible to fever blisters caused by sun exposure. The lips should be covered with a total-blocking cream that resists washing, sweating, and licking. Zinc oxide and lip balms containing PABA are both good. Reapply lip protection frequently.

INSECT REPELLENT

The wilderness is an occasional home for people but the permanent habitat of insects. Some of them—mosquitoes, biting flies, "no-see-um" gnats, ticks, chiggers—want to feast on your body. You can protect your body and blood with heavy clothing, including gloves and head nets in really buggy areas.

In hot weather, covering your body with heavy clothing may be unbearable, and insect repellents become a good alternative. Repellents with N,N-diethyl-metatoluamide (DEET) claim to be effective against all these insects, but are really best for keeping mosquitoes at bay. One application of a repellent with a high concentration of DEET will keep mosquitoes from biting for several hours, though they will still hover about. Mosquito repellents are marketed under many names, have differing potencies, and come in liquid, cream, spray, and stick form. Be aware that DEET can discolor or dissolve plastics, paints, and synthetic fabrics.

Despite claims by manufacturers, DEET is not very effective at repelling biting flies. Products with ethyl-hexanediol and dimethyl phthalate are much more effective against black flies, deer flies, and gnats. Unfortunately fly repellents don't do much to ward off mosquitoes.

Ticks are a potential health hazard due to the diseases they can carry, such as Lyme disease and Rocky Mountain spotted fever. In tick country, check clothing and hair frequently during the day. At night, give your clothes and body a thorough inspection to locate ticks before they embed themselves. If you find a tick on your body, cover it with heavy oil (mineral, salad, etc.) to close its breathing pores. The tick may disengage at once. If not, allow the oil to remain in place for half an hour, then carefully remove the tick with tweezers, taking care to remove all body parts. Once a tick is deeply embedded, you may need a physician to remove its buried head.

REPAIR KIT

It's helpful to carry a repair kit for your equipment. It could include wire, tape, safety pins, thread, needles, yarn, string, patches, and small pliers, and will probably grow over time as you add items you wished you had taken on a previous trip.

ICE AXE

An ice axe is indispensable on snowfields and glaciers and on snow-covered alpine trails in spring and early summer. It also has great value when you are traveling in steep heather, scree, or brush, crossing streams, and digging sanitation holes. However, an axe is more dangerous than helpful in unpracticed hands. Get instruction in its use.

EQUIPMENT CHECK LIST

Even the experienced can forget an important item in the rush to get ready for the next trip. Seasoned hikers and climbers use a check list as the only sure way to avoid an oversight. The following list is a good foundation for formulating your own personal check list. Add or subtract from this list as you see fit, then get in the habit of checking your own list before each trip.

Items in parentheses are optional, depending on your own preference and the nature of the trip.

Items marked with an asterisk can be shared by the group.

ALL TRIPS

THE TEN ESSENTIALS

1. Flashlight/headlamp, with spare bulbs and batteries
2. Map
3. Compass
4. Extra food
5. Extra clothing
6. Sunglasses
7. First-aid supplies
8. Pocket knife
9. Matches, in waterproof container
10. Fire starter

CLOTHING

Boots
Socks: inner and outer
Long underwear
Pants: wool or pile
Sweater or shirt: wool or pile
Parka: wind and rain
Hats: wool, rain, sun
Mittens, gloves
Down or synthetic garments
Wind/rain pants
(Gaiters)
(Shorts)
(Hot-weather shirt)

OTHER

Rucksack
Ice axe
Emergency shelter or bivy sack
Water bottle/purification tablets
Lunch
Sunscreen and lip protection
Bandanna
Toilet paper
Whistle
(Cup)
(Insect repellent)
(Moleskin)
(Nylon cord)
(Altimeter)
(Watch)
(Camera and film)
(Binoculars)
(Insulated seat pad)

ADDITIONAL ITEMS FOR OVERNIGHT TRIPS

Internal- or external-frame pack
Sleeping bag and stuff sack
Sleeping pad
Spoon
*Shelter: tent or tarp

---

* = equipment shared by the group
() = optional items

25

*Food
*Water container
*Repair kit
*Stove, fuel, and accessories
*Pots and cleaning pad
(Fork)
(Toiletries)
(Spare clothing)
(Camp footwear)
(Pack cover)

ADDITIONAL GEAR FOR GLACIER OR WINTER CLIMBS

Crampons
Carabiners
Slings: chest and prusik
Seat harness
Additional warm clothing: mittens, mitten
    shells, socks, balaclava, down or synthetic
    clothing, long underwear
Extra goggles
(Helmet)
(Supergaiters)
(Snowshoes or skis)
(Candles)
(Avalanche beacon)
*Climbing rope
*Rescue pulley
*Snow shovel
*Group first aid

26

*Alarm clock or alarm watch
*(Flukes, pickets, ice screws)
*(Wands)
*(Snow saw)

ADDITIONAL GEAR FOR ROCK CLIMBS

Helmet
Carabiners
Belay gloves
Runners
Seat harness
Prusik sling
*Climbing rope
*Rack: chocks, stoppers, etc.
*Chock pick
(Belay/rappel device)

THE EIGHT PRINCIPLES
OF WILDERNESS USE

To minimize impact on the wilderness environment, follow these principles when traveling in the outdoors.

1. Stay on established trails; do not cut switchbacks. When traveling cross-country, tread lightly to minimize damage to vegetation and soil slopes.

2. Camp in established campsites whenever available. Do not camp in fragile meadows. Camp on snow or rock when away from established campsites.

3. Use a camp stove instead of building a fire.

4. Properly dispose of human waste away from water, trails, and campsites.

5. Wash well away from camps and water sources. Properly dispose of waste water; avoid the use of non-biodegradable soap.

6. Pack out all party litter plus a share of that left by other parties.

7. Leave flowers, rocks, and other natural features undisturbed.

8. Keep wildlife healthy and self-reliant by not feeding them. Pack out all uneaten food. Leave pets at home.

SAFETY AND LEADERSHIP

Backcountry travel and mountaineering are sports of controlled risk. Hikers and climbers deal with the hazards of nature and their own short-comings by cultivating knowledge, skill, and good judgment. A list of dangers in the wilderness reads like a catalog of disaster unless you've studied the hazards and know how to avoid them in order to establish control over your own safety. Both hikers and climbers also need to develop leadership ability to help promote their group's safety, comfort, and success.

ACCIDENTS

One way to prevent accidents is to study incidents that have happened and try to learn from them. For climbers, the American Alpine Club does just that in its annual publication *Accidents in North American Mountaineering*. For climbers, the publication includes only actual climbing accidents, as distinguished from other mishaps that occur in mountainous regions. It describes only those accidents that are voluntarily reported, so it doesn't include numerous unpublicized incidents. The accidents represent only a fraction of those in mountaineering areas throughout the world.

More than 1,100 climbers have been killed in North American mountaineering accidents since 1947, when the American Alpine Club began its annual reports. Despite advances in equipment, skills, and techniques, a number of climbers are killed or injured each year. Every accident is a little different from every other, but there are many common causes and most involve human error. For example, modern gear rarely breaks by itself. Equipment failure is usually related to improper use or to poor placement of protection. Climbers also cause accidents when they try to exceed their climbing abilities by relying on incompletely learned techniques.

There are some limitations to the American Alpine Club statistics because the types of causes change as climbing practices evolve, and the classification of accidents is a difficult judgment call, especially in deciding the relationship of immediate and contributing causes. But in spite of these limitations, the annual reports roughly indicate the elements of danger in climbing by showing recurring patterns.

The most common immediate causes of accidents reported were (1) fall or slip on rock, (2) slip on snow and ice, and (3) falling rock or other object. The most common contributing causes were (1) climbing unroped, (2) attempting a climb that exceeded abilities (inexperience), and (3) being inadequately equipped for the conditions or climbing situation.

30

An accident victim may feel that a mishap occurred "like a bolt out of the blue." But it's clear, in retrospect, that many accidents moved step by step toward an almost predictable incident. The climber who takes off on an ascent without an ice axe because it was forgotten in the rush of leaving home may do fine all day in soft snow. If this climber slips during the descent on the firmer snow of evening and is injured, it's an accident—but it's no "bolt out of the blue." An alert leader tries to spot potential accidents, and takes such precautions as turning back a minimally equipped party in the face of worsening weather.

WHEN AN ACCIDENT HAPPENS

Leadership and discipline are key elements of success in alpine rescue. If your party has no recognized leader, select one to take charge in an emergency. The leader should consider suggestions from other members of the party, but the leader's decisions must be accepted without argument.

Go to the aid of an injured person quickly—but move carefully to avoid further accidents. On difficult terrain, dispatch only one or two rescuers, who are on belay and pack all the needed aid and rescue gear. As you climb or rappel down to the subject, keep to one side to avoid the danger of knocking rocks onto the victim. Stay calm; frenzied activity only complicates the rescue.

Urgent first aid should be rendered as soon as possible: bleeding stopped, breathing restarted, shock relieved, fractures immobilized. If it becomes necessary to move the victim to another spot because the accident site is too hazardous, use methods that will not compound the injuries.

It's important to keep close watch over the injured person. In climbing situations, a victim who is left unguarded, even for a few moments, must be tied to the mountain to prevent the danger of falling or wandering off, perhaps due to confusion or irrationality. Design the tie-in so that the climber cannot unfasten it.

In dealing with an emergency, swift action may be less effective than correct action. A hasty splitting up of the party or any other spontaneous but shortsighted effort has less chance of ultimate success than one more deliberately considered. Therefore, after the initial demands of safety and first aid are satisfied, the leader and party sit down to plan. Everything must be thought through to the very end, everything prearranged, including what each person is to do under all circumstances throughout the rescue operation.

Every aspect of the situation needs cool analysis. How serious are the injuries? What measures are necessary to sustain the victim during evacuation? What is the terrain like and how far is it to the road? What are the strengths and resources of the party? Only after careful analysis of such questions should your party select a course of action.

FIRST AID REPORT FORM

© The Mountaineers

START HERE			FINDINGS	FIRST AID GIVEN
AIRWAY, BREATHING, CIRCULATION INITIAL RAPID CHECK (Chest Wounds, Severe Bleeding)				
ASK WHAT HAPPENED: ASK WHERE IT HURTS:				
TAKE PULSE AND RESPIRATION: Pulse Respiration				
HEAD-TO-TOE EXAMINATION	HEAD:	Scalp—Wounds Ears, Nose—Fluid Eyes—Pupils Jaw—Stability Mouth—Wounds		
	NECK:	Wounds, Deformity		
	CHEST:	Movement, Symmetry		
	ABDOMEN:	Wounds, Rigidity		
	PELVIS:	Stability		
	Extremities:	Wounds, Deformity Sensation & Movement Pulse Below Injury		
	BACK:	Wounds, Deformity		
	SKIN:	Color Temperature Moistness		
STATE OF CONSCIOUSNESS				
PAIN (Location)				
LOOK FOR MEDICAL ID TAG				
ALLERGIES				
VICTIM'S NAME Age:				
COMPLETED BY Date/Time:				

KEEP THIS SECTION WITH VICTIM

RESCUE REQUEST
Fill Out One Form Per Victim • Send Out With Request For Aid

TIME OF INCIDENT   AM   PM   DATE

NATURE OF INCIDENT
FALL ON:
☐ ROCK   ☐ SNOW   ☐ FALLING ROCK   ☐ CREVASSE   ☐ AVALANCHE
☐ ILLNESS   ☐ EXCESSIVE HEAT   ☐ EXCESSIVE COLD
BRIEF DESCRIPTION OF INCIDENT

INJURIES (List Most Severe First)	FIRST AID GIVEN
SKIN TEMP/COLOR	
STATE OF CONSCIOUSNESS	
PAIN (Location)	

RECORD	Initial			When Leave Scene
Time				
Pulse				
Respiration				

VICTIM'S NAME

ADDRESS

NOTIFY (Name)

RELATIONSHIP   PHONE

EXACT LOCATION (Include Marked Map If Possible)
QUADRANGLE: SECTION: AREA DESCRIPTION:

TERRAIN: ☐ GLACIER ☐ SNOW ☐ ROCK ☐ BRUSH ☐ TIMBER ☐ TRAIL
          ☐ FLAT   ☐ MODERATE   ☐ STEEP
ON-SITE PLANS: ☐ Will Stay Put   ☐ Will Evacuate To _____
Can Stay Overnight Safely: ☐ Yes   ☐ No
On-Site Equipment:
☐ Tent   ☐ Stove   ☐ Food   ☐ Ground Insulation   ☐ Flare   ☐ CB Radio

LOCAL WEATHER

EVACUATION   ☐ Carry-Out   ☐ Helicopter   ☐ Lowering   ☐ Raising

EQUIPMENT: ☐ Rigid Litter   ☐ Food   ☐ Water   ☐ Other

PARTY MEMBERS REMAINING: ☐ Beginners   ☐ Intermediate   ☐ Experienced

NAME   NOTIFY(Name)   Phone

NOTIFY:
IN NATIONAL PARK: Ranger   OUTSIDE NATIONAL PARK: Sheriff/County Police,
RCMP (Canada)

VITAL SIGNS RECORD
Keep This Section With The Victim

Record TIME	BREATHS		PULSE		PULSES BELOW INJURY	PUPILS	SKIN	STATE OF CONSCIOUS-NESS	OTHER
	Rate	Character	Rate	Character					
		Deep, Shallow, Noisy, Labored		Strong, Weak, Regular	Strong, Weak, Absent, Irregular	Equal Size, React To Light, Round	Color, Temper-ature, Moistness	Alert, Confused, Unresponsive	Pain, Anxiety, Thirst, etc.